Anisotropic wave train events and interplanetary disturbances

Anisotropic wave trains

Rajesh Kumar Mishra
Rekha Agarwal Mishra

GRIN

Bibliographic information published by the German National Library:

The German National Library lists this publication in the National Bibliography; detailed bibliographic data are available on the Internet at http://dnb.dnb.de.

ISBN: 9783346988508
This book is also available as an ebook.

Print and binding: Books on Demand GmbH, Norderstedt, Germany
Printed on acid-free paper from responsible sources.

GRIN web shop: https://www.grin.com/document/1434252

Anisotropic wave train events and interplanetary disturbances

Rajesh K. Mishra and Rekha Agarwal Mishra

The name of the author to whom offprint requests should be sent:

Rajesh K. Mishra

Corresponding Author:

Rajesh K. Mishra

Abstract

We studied the cosmic ray intensity variation due to interplanetary magnetic clouds during low amplitude anisotropic wave train events. The low amplitude anisotropic wave train events in cosmic ray intensity using the data of ground based Deep River neutron monitor has been identified and studied during the period 1981-1994. It is observed that solar wind velocity remains higher than normal (> 300 km/s) and interplanetary magnetic field strength B remains lower than normal on the arrival of magnetic cloud during low amplitude anisotropic wave train events. It is also noted that the proton density remains significantly low during high solar wind velocity, which is expected. The north south component of interplanetary magnetic field Bz turns southward prior to one day of the arrival of cloud and remains in the southward direction after the arrival of cloud. The cosmic ray intensity is found to increase with the increase of solar wind velocity. It is noteworthy that decrease in cosmic ray intensity start not at the onset of cloud but after few days. The cosmic ray intensity significantly increases before and after 90 hours of the arrival of the magnetic cloud and decreases gradually after the passage of the magnetic cloud. The north south component of IMF (Bz) significantly correlated with latitude angle (θ°) and Dst index, whereas the geomagnetic activity index

(Ap) is significantly anti-correlated decreases with latitude angle (θ°) and Dst index on the arrival of interplanetary magnetic cloud during LAE.

Keywords: cosmic ray, solar wind, interplanetary magnetic field, magnetic cloud and geomagnetic activity.

PACS Nos.: 96.40.Kk, 96.40. -z, 96.40.cd

Introduction

Previously a lot of low amplitude anisotropic wave train events have been observed with a significant shift in the diurnal time of maximum to co-rotational direction or later hours [1 and references therein]. Agrawal and Bercovitch [2] have shown that the direction of the 22-year component is perpendicular to the diurnal anisotropy vector and is along the line 162° east of the Sun-Earh line; they have attributed the 11-year component to the variation of cut-off rigidity. A significant increase is observed in the amplitude of first three harmonics (diurnal/semi-diurnal/tri-diurnal) during the passage of high speed solar wind stream, whereas the direction of the anisotropy have no time variation characteristics associated with solar wind velocity and north south component of interplanetary magnetic field for three neutron monitoring stations located at different geomagnetic cutoff rigidities and altitudes [3].

Interplanetary magnetic clouds belong to one of the several classes of transient flows in the solar wind. Magnetic clouds as ideal force free objects (cylinders or spheres) are ejected near the Sun and followed beyond the Earths orbit. It is found that the decrease in cosmic ray intensity, which are associated with magnetic cloud preceded by a shock, are very high and these decrease starts few days earlier than the arrival of cloud at Earth. From the study of the time profile of these decrease, it is found that the onset time of a forbush type decrease produced by a shock associated cloud starts nearly at the time of arrival of the shock front at the Earth [4, 5] and the recovery is almost complete within a week. Forbush decreases associated with shock-associated cloud are caused by magnetic field variations associated with interplanetary disturbances [5].

Badruddin et al. [6] have reported a possible correlation between magnetic clouds and cosmic ray intensity decrease, while Kudo et al. [7] have reported an increases in cosmic ray intensity that may be related to the geomagnetic D_{st} index and Iucci et al. [8] have found short term increase in cosmic ray intensity occurring inside the Forbush decrease, that possibly may be associated with magnetic clouds. Zhang and Burlaga [9] infer that the cosmic rays are mainly modulated by fluctuation rather than by drifting in the strong smooth field in the magnetic cloud.

Intense interplanetary magnetic fields (IMFs) are of basic importance to solar wind physics, magnetospheric physics and cosmic ray physics and play a crucial role in the modulation of galactic cosmic rays [10]. Barouch and Burlaga [11] found that the individual magnetic enhancements are generally associated with depressions in cosmic ray intensity and Barouch and Sari [12] have further demonstrated that these depressions are not related to turbulence or random motions in the field and only the large-scale features of IMFs are important. However, Nishida [13] has emphasized the importance of scattering by turbulent magnetic field in producing the transient modulation.

The existence of unusual magnetized clouds of plasma emitted by the active Sun was proposed by Morrison [14] as a cause of worldwide decreases in cosmic ray intensity lasting for days and correlated roughly with geomagnetic storms. Klien and Burlaga [15] have defined a magnetic cloud as a structure of radial dimension ~ 0.25 AU (at 1 AU) in which the magnetic field strength is higher than average and the field direction changes nearly monotonically from large southern (northern) to large northern (southern) directions. The field geometry in such a magnetic cloud is consistent with a magnetic loop [16]. Burlaga and Klein [17] have discussed two magnetic clouds one of which was associated with an unusual cosmic ray depression while no appreciable depressions in cosmic ray intensity was found in association with the other.

An attempt has been made in this paper to investigate the likely reason causing the occurrence of these types of unusual events in cosmic-ray intensity observed over the period 1981 – 1994.

Data Analysis

The pressure-corrected data of the Deep River Neutron Monitor (NM) station (data from http://spidr.ngdc.noaa.gov/NeutronMonitor) has been subjected to Fourier analysis for the period 1981 – 1994 after detrending. While performing the analysis of the data, all those days discarded having more than three continuous hours of data missing.

Using the long-term plots of the cosmic-ray intensity data as well as the amplitude observed from the cosmic-ray pressure-corrected hourly neutron monitor data using harmonic analysis, the low-amplitude wave train events (LAE) have been selected on the basis of following criteria:

- Low-amplitude wave train events of continuous days have been selected when the amplitude of the diurnal anisotropy remains lower than 0.3% on each day of the event for at least five days.

- In the selection of these events, special care has been taken, *i.e.* if there occurred any pre-Forbush decreases or post-Forbush decrease before or after the event or the event is in the recovery phase or declining phase they are not considered.

On the basis of the above selection criteria we have selected 28 LAEs during the period 1981 – 1994. The hourly cosmic-ray intensity data for Deep River NM station [Geog. Lat. 46.10°, Geog. Long. 282.50°, Vertical cut-off rigidity 1.02 (GV)] have been investigated in the present study. Figure 1 shows examples some of the low amplitude anisotropic wave trains.

In the present study we have identified the interplanetary magnetic clouds using plots of hourly values of interplanetary parameters [18-22] to study the role of these clouds in LAE. The large Forbush decreases in cosmic-ray intensity, if any, have been excluded to avoid their influence.

We have adopted the chree analysis of superposed epoch to study the effect of interplanetary magnetic clouds on cosmic-ray intensity using the hourly-average cosmic-ray intensity of the Deep River neutron monitor during the passage of LAEs.

Further, various features, which are observed over the solar disk (data from http://nssdc.gsfc.nasa.gov/omniweb) during the periods of events, have also been studied.

Results and Discussion

The existence of the interplanetary-shock-associated magnetic clouds [16, 17] has provided a new tool for investigating the physical process responsible for cosmic-ray decreases. Many authors [23-25] using the superposed-epoch-analysis technique for a number of events have demonstrated that the turbulent sheath between the interplanetary shock and the magnetic cloud essentially produces the decreases. Sanderson et al. [26,

27] using the magnetic cloud data of Marsden et al. [28] have clearly shown that magnetic clouds also produce cosmic-ray decreases. They have suggested that post-shock regions, tangential discontinuities, and magnetic clouds are equally effective in producing cosmic-ray decreases. Kahler and Reames [29], considering magnetic fields of different topology, have suggested that the cosmic-ray decreases could also be produced by the passage of magnetic clouds with open field-line configurations. Using the methodology of Zhang and Burlaga [9] we have identified positive and negative magnetic clouds in 4 LAEs out of 28 LAEs during these events. These magnetic clouds have been divided into two categories, namely those associated with shocks and those not associated with shocks. Some of them are negative clouds without shocks and other positive clouds with and without shock.

The cosmic ray intensity, IMF and solar wind plasma parameters alongwith Dst index have been plotted in figure 2 to show an example of interplanetary positive magnetic cloud (magnetic field is directed northward) without shock occurred on April 20, 1981 at 2300 UT during these events. It is clearly seen from figure 2 that cosmic ray intensity is found to increase prior to arrival of cloud upto the passage of the cloud. The north south component of IMF (Bz) increases prior to onset of cloud upto the onset of cloud and then remains statistically constant during the passage of cloud. It is also noteworthy that the north south component Bz significantly remains negative only during the passage of cloud. The solar wind velocity remains statistically constant prior to the arrival of cloud and start increasing on the arrival of cloud for few days then decreases gradually. The IMF strength (B) is slightly increases prior to the cloud then decreases slightly after the arrival of cloud. The disturbance storm time index Dst significantly increases prior to the arrival of cloud and then increases gradually during the passage of cloud. The proton density N is found to remain statistically constant before and after the arrival of cloud. The geomagnetic activity index (Ap) reaches to its maximum (121 nT), 7 days prior to the onset of magnetic cloud and then decreases sharply and reaches to its usual value (25 nT). The Ap index increases slightly 2 days prior to magnetic cloud event and then remains statistically constant with some deviations later on. One can clearly see from the shape of the plots that proton density (N) and Ap index shows similar deviations with a time lag of one day. The solar activity represented by sunspot numbers (R) is found remains constant before and after the onset of interplanetary magnetic cloud event. The latitude angle (θ°) increases with some deviations upto the arrival of cloud and then found to remain constant for few days during the passage of cloud. The longitude angle (ϕ°) increase few days prior to the onset of cloud upto 2 days after the cloud then significantly decrease for one day and then remain constant for few days. We can see from the plot that solar wind velocity (V) remains higher than normal (> 300 km/s) and IMF strength B remains lower than normal during this period. It is also evident that the proton density (N) remains significantly low during high solar wind velocity, which is expected. The north south component of IMF (Bz) turns southward prior to one day of the arrival of cloud and remains in the southward direction after the arrival of cloud. The cosmic ray intensity is found to increase with the increase of solar wind velocity.

To further find out a possible correlation between these parameters on the arrival of magnetic cloud, we have plotted scattered plots (for significant and good correlation only) alongwith regression equation in Fig 3 (a-j) and calculated the correlation coefficient between them and shown in Table 1. As depicted in Fig 3 a and Table 1 the cosmic ray intensity increases with increase of disturbance time index (Dst) and show a good positive correlation (r = 0.65). The north south component of IMF (Bz) is significantly increases with the increases of latitude angle (θ°) and Dst index and shows significant positive correlation with θ° (r = 0.86) and Dst index (r = 0.80). The geomagnetic activity index (Ap) as depicted in Fig 3 (d-f) significantly decreases with the increase of Dst index and latitude angle θ° showing significant anti-correlation (r = -0.84, r = -0.83). However, Ap is found to significantly increase with the increase of IMF strength (B) and shows positive correlation with B (r = 0.70). The sunspot numbers (R) seems to increase with the increase of IMF B and shows a good positive correlation with IMF B (r = 0.54). As seen in Fig 3 (h, i) the latitude angle θ° observed to increase with the increase of disturbance storm time index Dst and shows a significant positive correlation (r = 0.80) and cosmic ray intensity increases gradually with the increase of Dst index and shows a good positive correlation (r = 0.65). One can see from the Fig 3j that the proton density (N) increases with the increase of IMF strength (B) and shows a significant positive correlation (r = 0.84). The remaining parameters have no significant correlation with each other on the arrival of magnetic cloud during LAE.

To study the effect of these magnetic clouds on cosmic ray intensity during LAE, We have adopted the Chree analysis of superposed epoch for hours –141 to +121 and plotted in Fig 4 as a percent deviation of cosmic ray hourly intensity data alongwith statistical error bars (I) for Deep River during the period 1981-94. Deviation for each event is obtained from the overall average of 265 hours. Epoch hour (zero hour) correspond to the starting hours of interplanetary magnetic cloud. One can see from figure 3 that the cosmic ray intensity significantly enhanced 100 hours prior to the onset of magnetic cloud event and reaches to its maximum in 10 hour time period on -89 hours and then decreases sharply. The cosmic ray intensity reaches to its usual value on 26th hour before the arrival of magnetic cloud. Significant deviations are evident from the figure before and after the onset of magnetic cloud. The cosmic ray intensity once again found to significantly enhanced 91 hour after the arrival of magnetic cloud and reaches to its maximum in 10 hour time period on +104 hour. The cosmic ray intensity is also reaches to its peak 15 hours after on –119 hour on the arrival of cloud. Thus we can see from this plot that decrease in cosmic ray intensity start not at the onset of cloud but after few hours. The cosmic ray intensity significantly increases before and after 90 hours of the arrival of the magnetic cloud and decreases gradually after the passage of the magnetic cloud. This is in good agreement with earlier findings reported by Yadav et al. [30]. These observations suggest that the cosmic-ray increase is essentially triggered by the passage of a magnetic cloud during LAEs. Sanderson *et al.* [26, 27] and Kahler and Reames [29] found that magnetic clouds are effective in producing cosmic-ray decreases. Interplanetary disturbances (Magnetic cloud) are found to be the responsible factor in producing the decrease in cosmic-ray intensity on a short-term basis [31]. Mishra et al. [32] concluded from their analysis that magnetic clouds in association with sudden storm commencement (SSC) events produce large decreases in cosmic-ray intensity start not on the onset of cloud but after 2 days.

The decrease occurs in cosmic-ray intensity associated with clouds preceded by shocks not at the arrival time of the clouds but earlier. The onset of these decreases as observed on the Earth is almost coincident with the shock arrival at the Earth. A turbulent sheath of ambient plasma in which there may be large fluctuations in both the strength and direction of the magnetic field follow the shocks. This in turn is followed by the mass ejecta, which is driving the shock. The shock and associated ejecta both are important in determining the time profile of the decrease. The field magnitude and the speed of this modulating region both seem to be related to the magnitude of cosmic-ray decreases.

Conclusions

On the basis of the present investigation the following conclusions have emerged:

- The cosmic ray intensity significantly increases before and after 90 hours of the arrival of the magnetic cloud and decreases gradually after the passage of the magnetic cloud.
- The north south component of interplanetary magnetic field Bz turns southward prior to one day of the arrival of cloud and remains in the southward direction after the arrival of cloud.
- The solar wind velocity remains higher (> 300) than normal and interplanetary magnetic field strength B remains lower than normal on the arrival of magnetic cloud events for low amplitude anisotropic wave trains.
- The proton density remains significantly low during high solar wind velocity during the passage of interplanetary magnetic clouds.
- The north south component of IMF (Bz) significantly correlated with latitude angle (θ°) and Dst index, whereas the geomagnetic activity index (Ap) is significantly anti-correlated decreases with latitude angle (θ°) and Dst index on the arrival of interplanetary magnetic cloud during LAE.

Acknowledgements

The authors are indebted to various experimental groups, in particular, Prof. Margret D. Wilson, Prof. K. Nagashima, Miss. Aoi Inoue and Prof. J. H. King for providing the data. We also acknowledge the use of NSSDC OMNI database and NGDC geophysical data.

References

[1] Rajesh K Mishra, and Rekha Agarwal Mishra, *Ind. J. Radio and Space Phys.* **33**, 285 (2004).

[2] S P Agrawal, and M Bercovitch, *18th Int. Cosmic Ray Conf., Bangalore* **3,** 316 (1983).

[3] Rekha Agarwal Mishra, and Rajesh K. Mishra, *Astrophysics* **49 (4)**, 555, 2006.

[4] S P Duggal, M A Pomerantz, R K Schaefer, and C H Tsao, *J. Geophys. Res.* **88**, 2973 (1983).

[5] Badruddin, R S Yadav, and N R Yadav, *Solar Phys.* **105,** 413 (1986).

[6] Badruddin, R S Yadav, and S P Agrawal, *19th Int. Cosmic Ray Conf., 1-12 NASA Cof. Publ.* 5, **258** (1985).

[7] S Kudo, M Wada, P Tanskanen, and M Kodama, *19th Int. Cosmic Ray Conf., 1-8 NASA conf. Publ.* **5**, 246 (1985).

[8] N Iucci, M Parisi, C Signorini, M Storini, and G Villoresi, 19th Int. Cosmic Ray Conf. Publ. **5**, 226 (1985).

[9] Z Zhang, and L F Burlaga, *J. Geophys. Res.* **93**, 2511 (1988).

[10] L F Burlaga, and J H King, *J. Geophys. Res.* **84**, 6633 (1979).

[11] E Barouch, and L F Burlaga, *J. Geophys. Res.* **80**, 449 (1975).

[12] E Barouch, and J W Sari, *J. Geophys. Res.* **81**, 1453 (1976).

[13] A Nishida, *J. Geophys. Res.* **87**, 6003 (1982).

[14] P Morrison, *Phys. Rev.* **95**, 641 (1954).

[15] L W Klein, and L F Burlaga, *J. Geophys. Res.* **87,** 613 (1982).

[16] L F Burlaga, E Sittler, F Mariani, and R Schwenn, *J. Geophys. Res.* **86,** 6673 (1981).

[17] L F Burlaga, and L W Klein, *NASA Techn. Mem.* 80668 (1980).

[18] J King, *Interplanetary Medium Data Book-Supplement 2, NSSD/WDC-A,* Goddard Space Flight Center, Greenbelt, MA (1983).

[19] J King, *Interplanetary Medium Data Book-Supplement 3, NSSD/WDC-A,* Goddard Space Flight Center, Greenbelt, MA (1986a).

[20] J King, *Interplanetary Medium Data Book-Supplement 3A, NSSD/WDC-A,* Goddard Space Flight Center, Greenbelt, MA (1986b).

[21] J King, *Interplanetary Medium Data Book-Supplement 4, NSSD/WDC-A,* Goddard Space Flight Center, Greenbelt, MA (1989).

[22] J King, and N E Papitashvili, *Interplanetary Medium Data Book, NSSCD/WDC-A-R@S94-08,* Goddard Space Flight Center, Greenbelt, Maryland (1994).

[23] Badruddin, D Venkatesan, and B Y Zhu, *Solar Phys.* **134**, 203 (1991).

[24] Badruddin, R S Yadav, and N R Yadav, *Solar Phys.* **105**, 413 (1986).

[25] G Zhang, and L F Burlaga, *J. Geophys. Res.* **93**, 2511 (1988).

[26] T R Sanderson, J Beeck, R G Marsden, C Tranquille, K.P Wenzel, R B McKibben, and E J Smith, *21st Int. Cosmic ray Conf.* **6**, 251 (1990a).

[27] T R Sanderson, J Beeck, R G Marsden, C Tranquille, K P Wenzel, R B McKibben, and E J Smith, *21st Int. Cosmic ray Conf.* **6,** 255 (1990b).

[28] R G Marsden, T R Sanderson, C Tranquille, K P Wenzel, and E J Smith, *J. Geophys. Res.* **92**, 11009 (1987).

[29] S W Kahler, and D V Reames, *J. Geophys. Res.* **96**, 9419 (1991).

[30] R S Yadav, N R Yadav, and Badruddin, *20th Int. Cosmic Ray Conf.* **4**, 83 (1987).

[31] S C Kaushik, and P K Shrivastava, *Bull. Astr. Soc. India.* **27**, 85 (1999).

[32] M P Mishra, P K Shrivastava, and D P Tiwari, *29th Int. Cosmic Ray Conf.* **2**, 331 (2005).

Captions to Figures

Fig 1: Cosmic Ray intensity records (Neutron Monitor count rates) at Deep River showing the occurrence of some low amplitude anisotropic wave trains.

Fig 2: Daily variation of cosmic rays from -5 to +5 day for the magnetic cloud event of April 20, 1981 alongwith IMF (Bz), solar wind velocity (V), IMF (B), Dst index, proton density (N), proton temperatue (T), Latitude Angle and longitude Angle.

Fig 3: Cross correlation between cosmic ray intensity, north south component of IMF (Bz), Dst index, Ap index, sunspot numbers (R), IMF strength (B), proton density

(N), latitude angle (θ°) alongwith regression equation and correlation coefficient (R) on the arrival of magnetic cloud during LAE.

Fig 4: Superposed epoch results of cosmic Ray intensity at Deep River NM station due to interplanetary magnetic clouds alongwith statistical error bars (I) during low amplitude anisotropic wave trains.

Table 1: Correlation coefficient (r) between cosmic ray intensity and different solar/interplanetary parameters on the onset of magnetic cloud during the passage of low amplitude anisotropic wave trains.

	Bz (nT)	**SWV (km/s)**	**Ap (nT)**	**SSN**	**Dst (nT)**	**N (cm³)**	**T °K**	**B (nT)**	**CRI**	**θ°**
SWV (km/s)	-0.27									
Ap (nT)	-0.73	0.23								
R	0.13	-0.43	0.15							
Dst (nT)	**0.79**	-0.30	**-0.84**	-0.11						
N (cm³)	-0.13	-0.34	0.66	0.37	-0.32					
B (nT)	-0.19	-0.14	**0.70**	**0.54**	-0.41	**0.84**	-0.28			
CRI	0.32	-0.02	-0.34	-0.09	**0.65**	-0.07	0.01	-0.14		
θ°	**0.86**	-0.11	**-0.83**	-0.13	**0.77**	-0.44	0.23	-0.45	0.34	
ϕ°	-0.50	0.27	0.13	-0.45	-0.19	-0.29	0.11	-0.20	0.01	-0.19

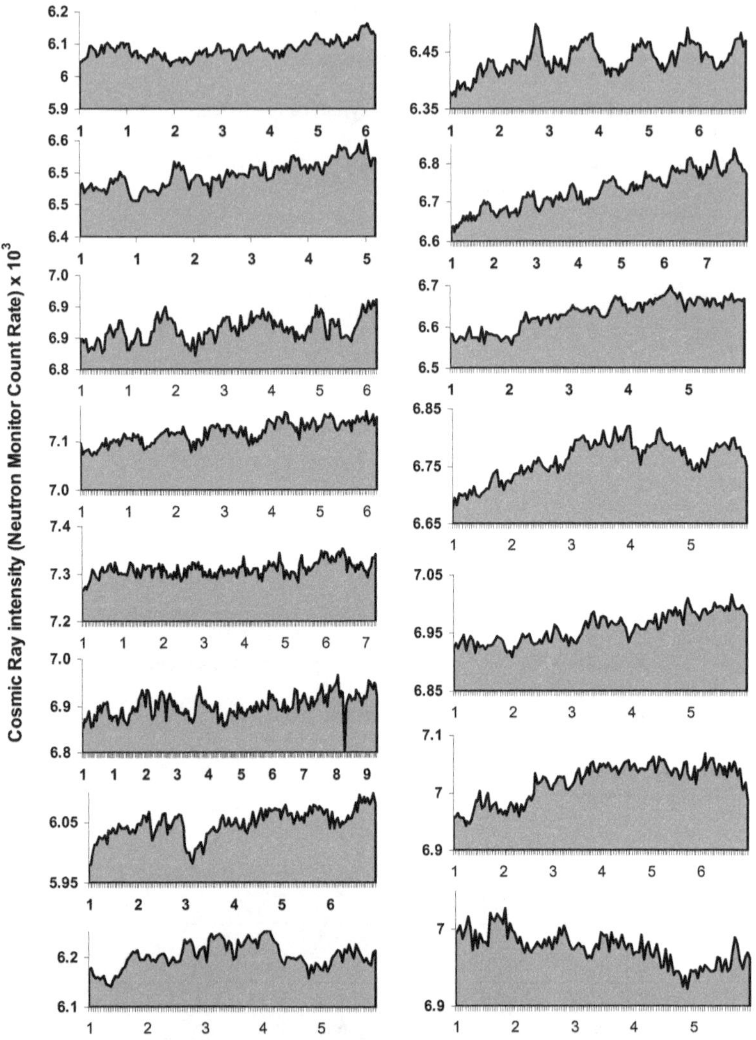
Cosmic Ray intensity (Neutron Monitor Count Rate) x 10^3

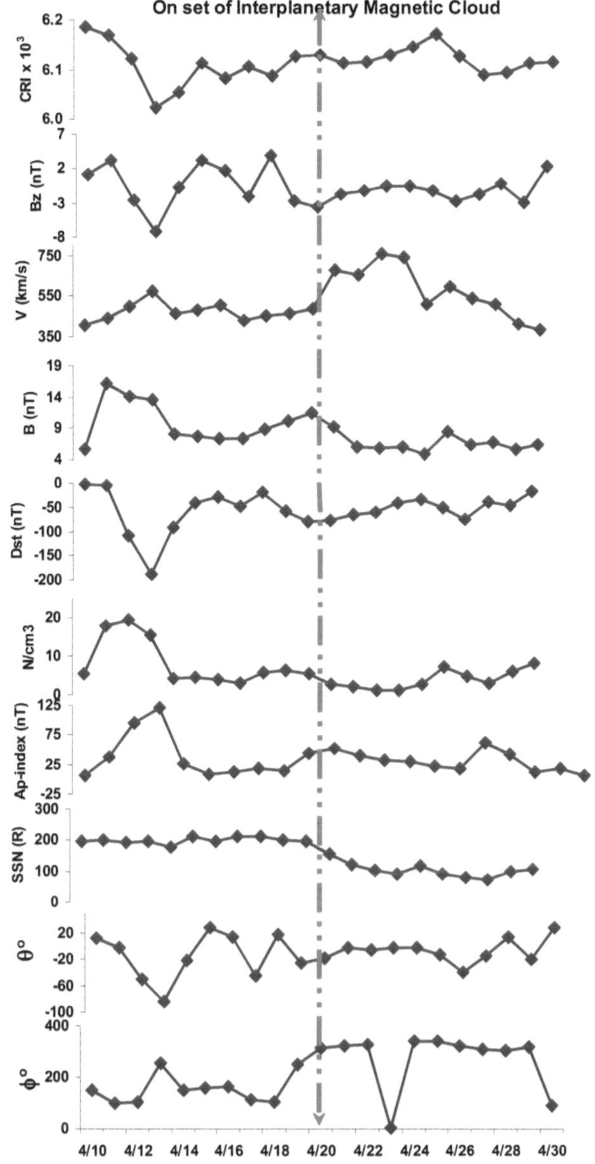

On set of Interplanetary Magnetic Cloud
CRI x 10^3
6.2
6.1
6.0
Bz (nT)
7
2
-3
-8
V (km/s)
750
550
350
B (nT)
19
14
9
4
Dst (nT)
0
-50
-100
-150
-200
N/cm3
20
10
0
Ap-index (nT)
125
75
25
-25
SSN (R)
300
200
100
0
θ°
20
-20
-60
-100
ϕ°
400
200
0
4/10
4/12
4/14
4/16
4/18
4/20
4/22
4/24
4/26
4/28
4/30

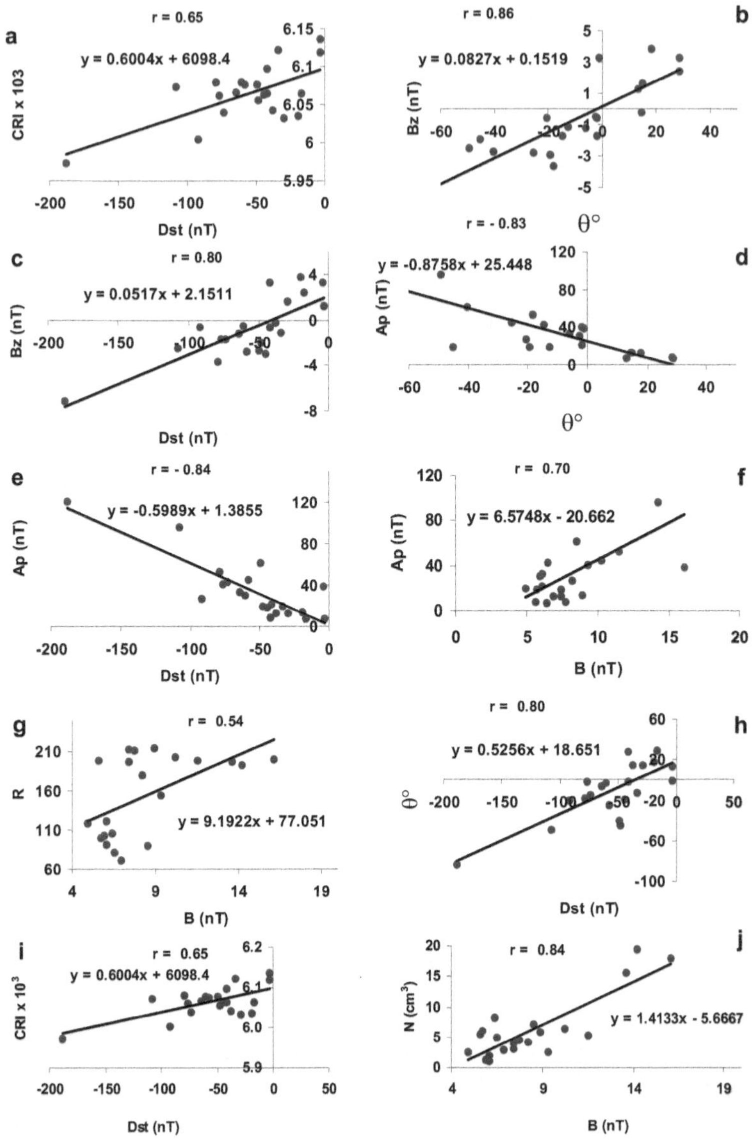

a
r = 0.65
y = 0.6004x + 6098.4
CRI x 103
Dst (nT)
b
r = 0.86
y = 0.0827x + 0.1519
Bz (nT)
θ°
c
r = 0.80
y = 0.0517x + 2.1511
Bz (nT)
Dst (nT)
d
r = - 0.83
y = -0.8758x + 25.448
Ap (nT)
θ°
e
r = - 0.84
y = -0.5989x + 1.3855
Ap (nT)
Dst (nT)
f
r = 0.70
y = 6.5748x - 20.662
Ap (nT)
B (nT)
g
r = 0.54
y = 9.1922x + 77.051
R
B (nT)
h
r = 0.80
y = 0.5256x + 18.651
θ°
Dst (nT)
i
r = 0.65
y = 0.6004x + 6098.4
CRI x 10^3
Dst (nT)
j
r = 0.84
y = 1.4133x - 5.6667
N (cm^3)
B (nT)

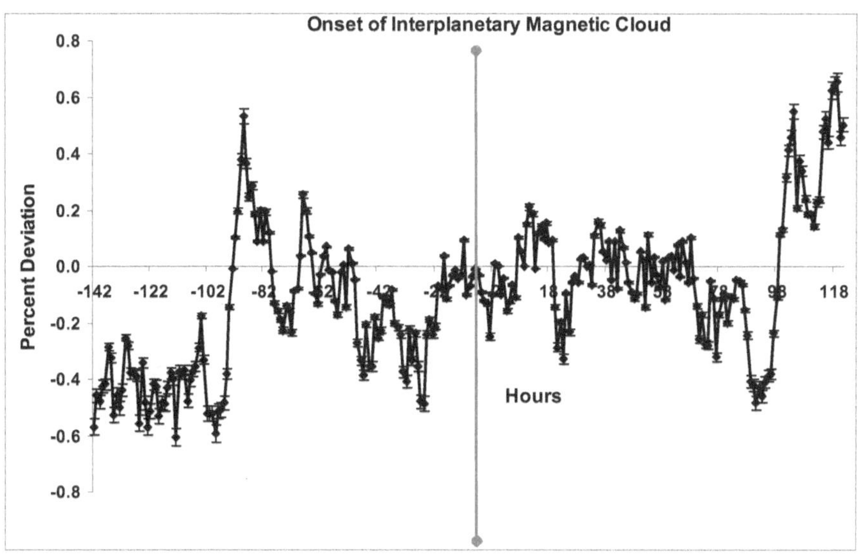

Onset of Interplanetary Magnetic Cloud
Percent Deviation
0.8
0.6
0.4
0.2
0.0
-0.2
-0.4
-0.6
-0.8
-142
-122
-102
118
Hours

www.ingramcontent.com/pod-product-compliance
Lightning Source LLC
La Vergne TN
LVHW042258190726
843491LV00016BA/2979

* 9 7 8 3 3 4 6 9 8 8 5 0 8 *